baby PARROTS

KIM THOMPSON

CREATIVE EDUCATION • CREATIVE PAPERBACKS

CONT

TENTS

I AM A CHICK.

I am a baby parrot.

I was born with no feathers. They will grow later.
eye
beak

My mom and dad found a cavity in a tall tree. They lined it with soft plants.

My mom laid an egg inside. My dad brought food to her. After one month, I hatched!

After three months, I have colorful feathers. I am ready to fly. My parents help me.

My hooked
beak can
crack nuts.

I am smart. I like to learn and play.

I make many different sounds. I can even imitate human words!

I may live as long as 80 years.

SPEAK AND LISTEN

BA

Can you speak like a chick?

Baby parrots squawk and scream.

Listen to these sounds:

https://www.youtube.com/watch?v=aj3ny_GTuhM

CHICK WORDS

cavity: a hollow place or opening

feathers: light, soft parts that cover a bird's body

hatched: broke out of an egg to be born

imitate: to copy what someone says or does

READING CORNER

Bodden, Valerie. *Parrots (Amazing Animals)*. Mankato, Minn.: Creative Paperbacks, 2022.

Eaton, Maxwell. *The Truth about Parrots*. New York: Roaring Brook Press, 2021.

Roth, Susan L., and Cindy Trumbore. *Parrots over Puerto Rico*. New York: Lee & Low Books, 2025.

INDEX

PUBLISHED BY CREATIVE EDUCATION AND CREATIVE PAPERBACKS
P.O. Box 227, Mankato, Minnesota 56002
Creative Education and Creative Paperbacks are imprints of The Creative Company
www.thecreativecompany.us

LIBRARY OF CONGRESS CATALOGING-IN-PUBLICATION DATA
Names: Thompson, Kim, 1970- author
Title: Baby parrots / Kim Thompson.
Description: Mankato, Minnesota : Creative Education and Creative Paperbacks, [2026] | Series: Starting out | Includes bibliographical references and index. | Audience term: juvenile | Audience: Ages 4-7
Creative Education and Creative Paperbacks | Audience: Grades K-1 Creative Education and Creative Paperbacks | Summary: "Introduce beginning readers to the world of baby parrots with this life science starter. Includes photos, a labeled animal diagram, "Make a Noise" section, glossary, and further resources"-- Provided by publisher.
Identifiers: LCCN 2024043255 (print) | LCCN 2024043256 (ebook) | ISBN 9798889897521 library binding | ISBN 9781682778388 paperback | ISBN 9798889897651 ebook
Subjects: LCSH: Parrots--Infancy--Juvenile literature
Classification: LCC QL696.P7 T486 2026 (print) | LCC QL696.P7 (ebook) | DDC 598.7/11392--dc23/eng/20250107
LC record available at https://lccn.loc.gov/2024043255
LC ebook record available at https://lccn.loc.gov/2024043256

DESIGN AND PRODUCTION
Design by Rhea Magaro
Production by Beeline Media and Design, Inc.
Art direction by Tom Morgan

PHOTOGRAPHS by Dreamstime/Tahreerqau, cover; Shutterstock/Bohbeh, 13, Eric Isselee, 4, 5, 11, David Ryo, 7, Ian Duffield, 8, PRIN6395, 10-11, Rosa Jay, 2-3, Wut_ Moppie, 6-7, yakub88, 9, YK, 12, ZoranOrcik, 14

Printed in India